DOGS
A Counting & Comparing Book

Copyright © Frances Mackay 2024
www.francesmackay.com
Design by Nicky Scott
www.nickyscottdesign.com
Illustrations licensed from Dreamstime
www.dreamstime.com
Main contributor Igor Zakowski

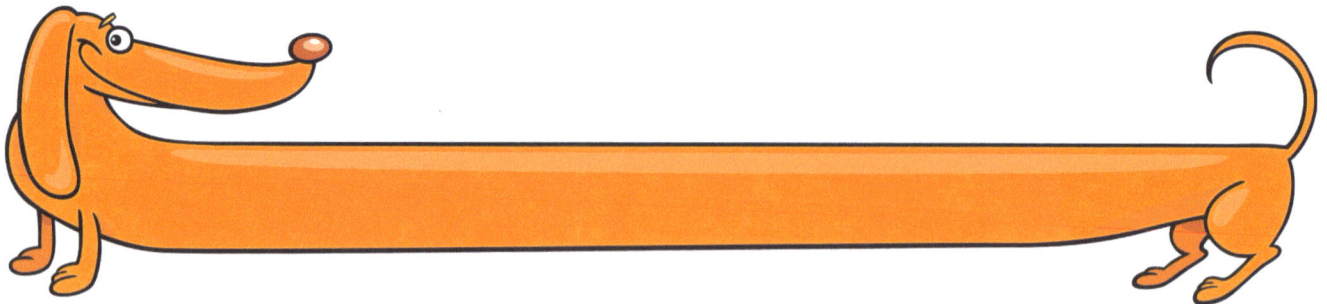

ISBN 978 0 646 890043

DOGS

A Counting & Comparing book

Frances Mackay

ONE sniffing dog.

What will puppy find?

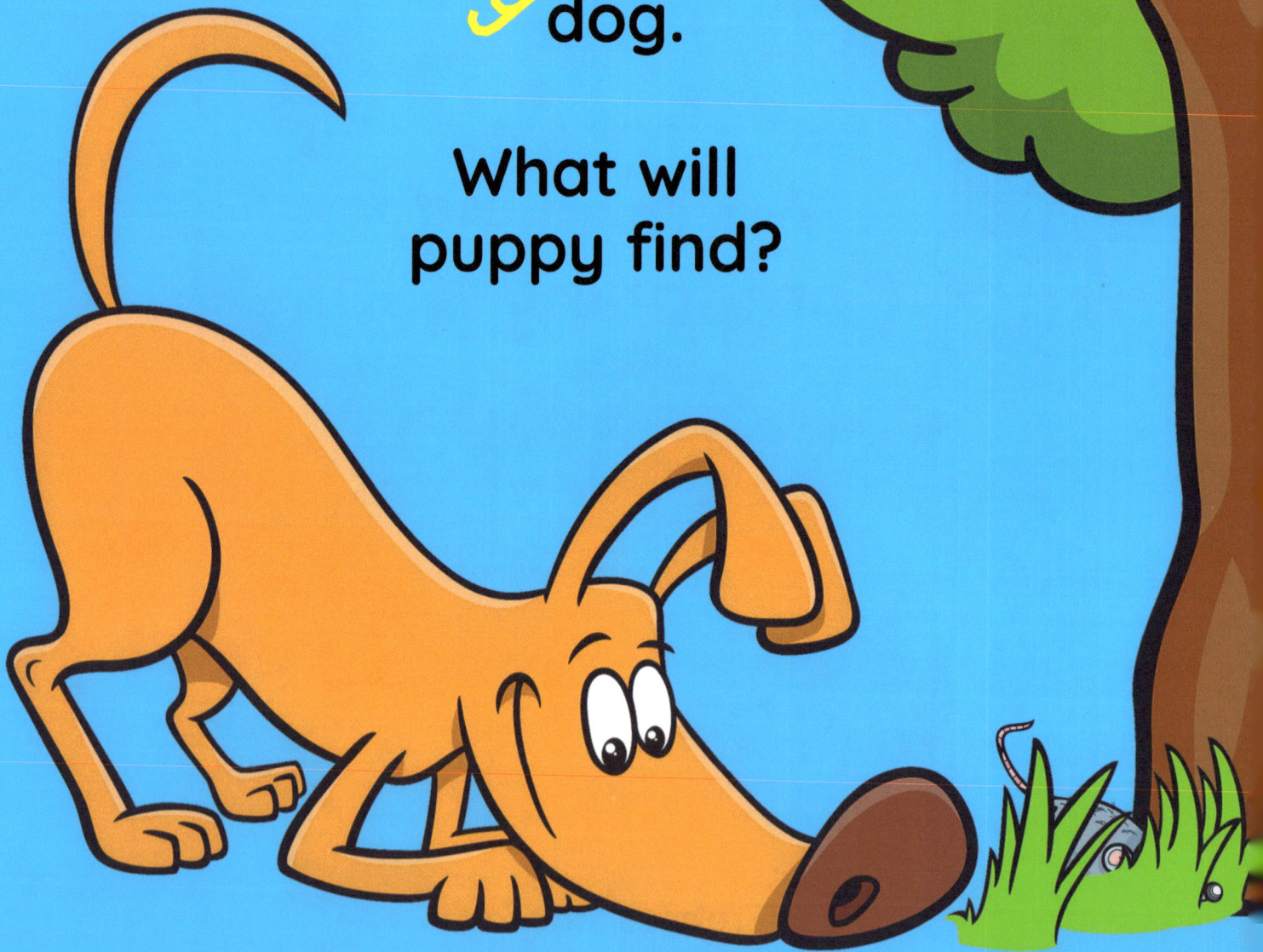

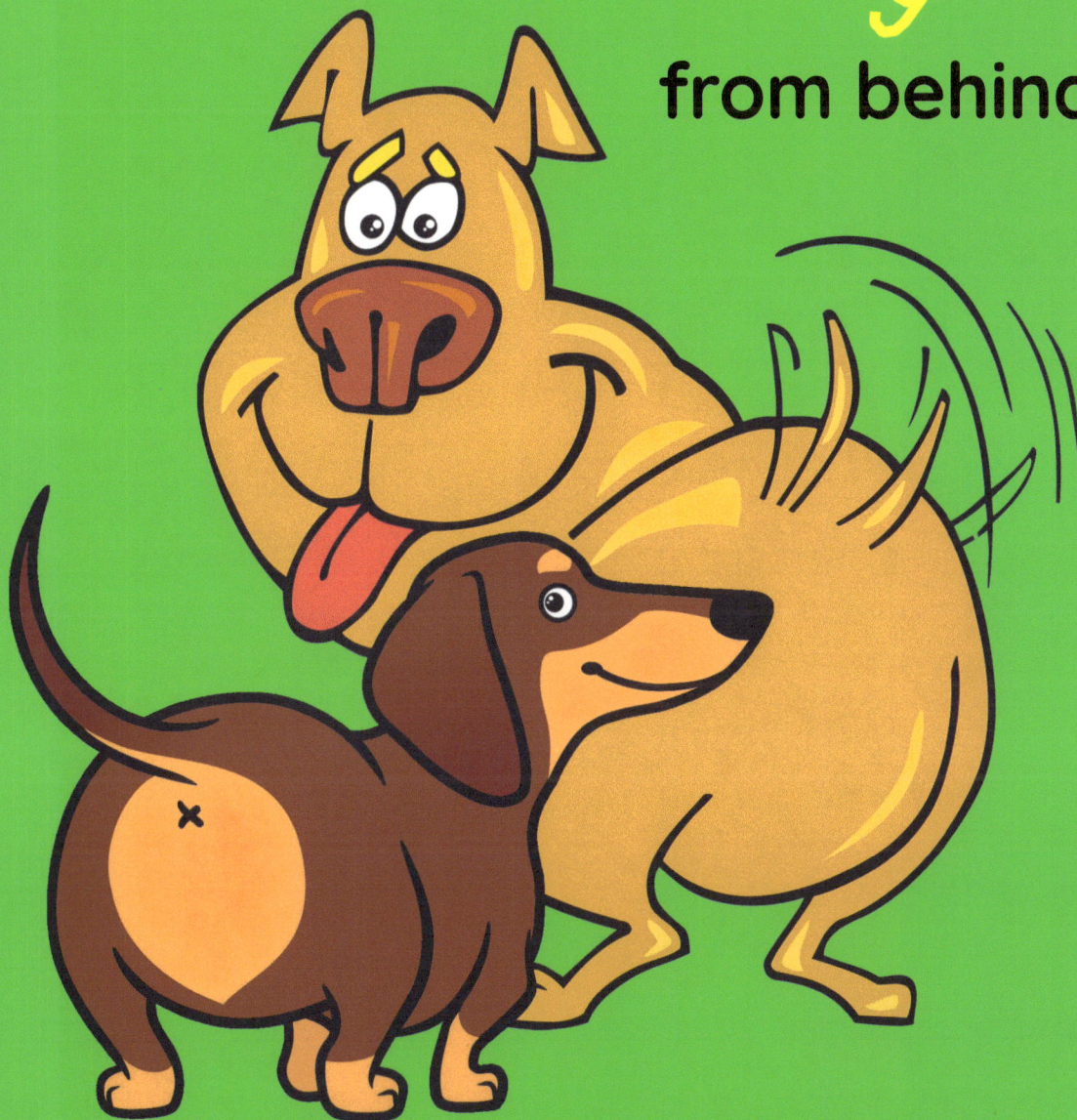

TWO doggy tails,

wagging

from behind.

THREE friendly dogs.
BIG. Medium. Small.

FOUR
pug
bottoms!

Find them
on the wall.

SIX quiet dogs dozing off at noon.

Tall dog.

Short dog.

A **Huge** dog asleep.

Look at this dog!
He's driving a jeep!

SEVEN dogs with long hair.

SEVEN dogs without.

Which of these **EIGHT** dogs has the longest snout?

NINE dashing dachs
hope to be the winner.

How many dogs
are running late for dinner?

Now that we've counted all the way to TEN,

let's count backwards
down to ONE again!

10 9 8 7

6 5 4 3

2 1

TEN relaxed dogs, all different sizes.

NINE sporty dogs, doing exercises.

EIGHT happy dogs.
They all want to play.

SEVEN sad dogs.
They've had a "ruff ruff" day.

SIX dogs with funny ears.

Droopy ears.

Pointy ears.

Waving-in-the-breeze ears.

Stick out ears.

Stick up ears.

Did-you-just-say-cheese? ears.

FIVE young dogs.

FOUR very old.

This old fella has
a story to be told.

Long dog.

Short dog.

One in between.

This **COSY** dog is treated like a queen!

THREE dogs with light fur

and TWO that are dark.

ONE *sniffing* dog.

He's still at the park!

What did he find?

Can you find these dogs in the book?

Collie

Scottish Terrier

Basset Hound

Wire Fox Terrier

Shar-Pei

Poodle

Schnauzer

St.Bernard

Afghan Hound

Bulldog

Bull Terrier

Dalmatian

German Shepherd

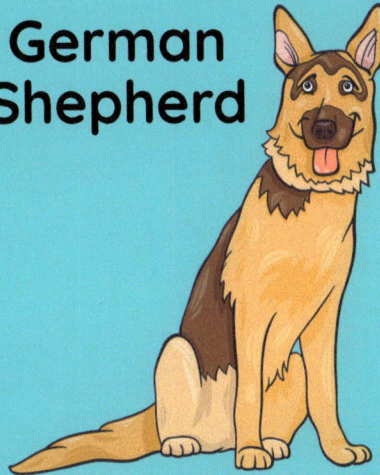

Pug

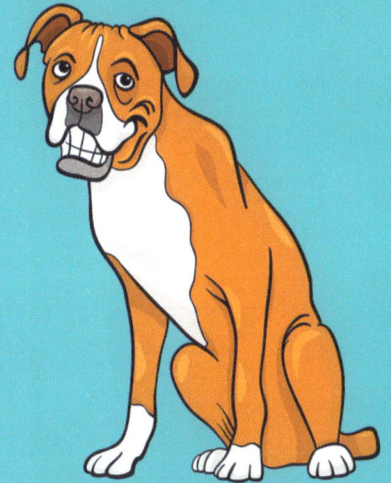

Boxer

Great Dane

Dachshund

Yorkshire Terrier

ABOUT THE AUTHOR - FRANCES MACKAY

Frances Mackay is the author of more than 90 teacher resource books and has written several picture books and activity books. She was a primary school teacher for 20 years in Australia and the UK. She has always loved animals and has shared her life with many cats, dogs, guinea pigs and other creatures over the years. Her favourite breed of dog is a Golden Retriever. She currently lives in Tasmania, Australia.

In memory of Gemma & Elsa

Read more about Frances and grab some FREEBIES at:
www.francesmackay.com

Did You Enjoy This Book?

Your feedback helps me provide the best quality books & helps other readers like you discover great books.

You can leave a review on Amazon or send it direct to me at:
frances@francesmackay.com

I read and appreciate each one. ❤️

Dogs Colouring Pack

Grab your FREE
10-page
Dogs Colouring Pack
when you visit!

THANK You! 😊

Books by Frances Mackay
www.francesmackay.com

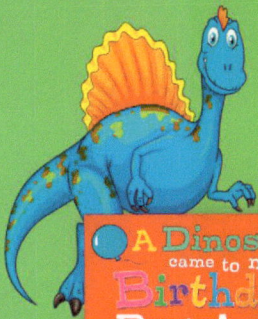

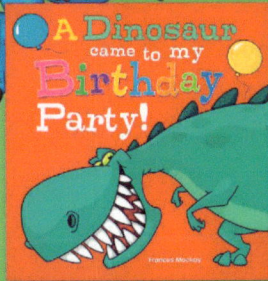

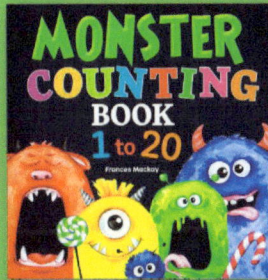

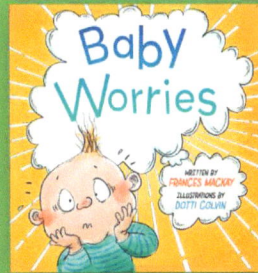

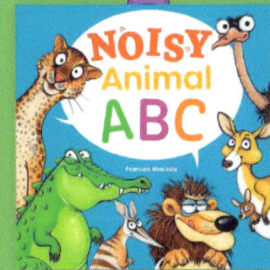

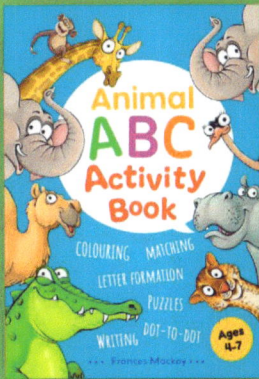

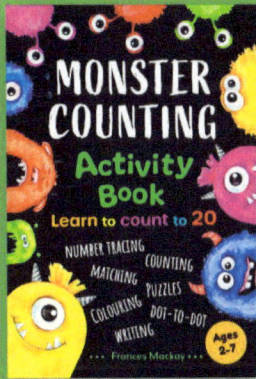

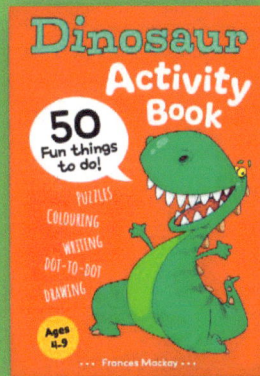

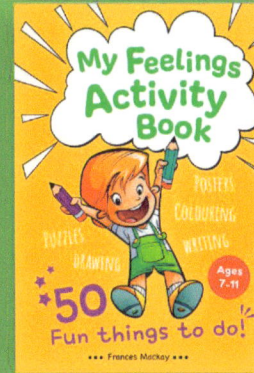

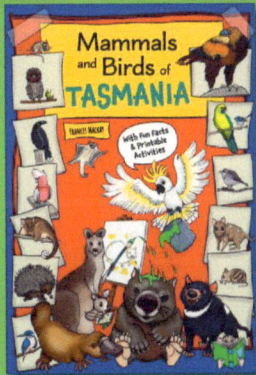

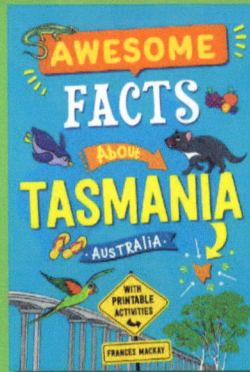

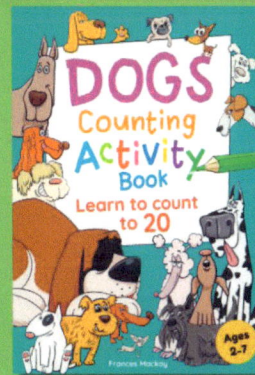

A Dinosaur came to my Birthday Party!

MONSTER COUNTING BOOK 1 to 20
Frances Mackay

Baby Worries
WRITTEN BY FRANCES MACKAY
ILLUSTRATIONS BY DOTTI COLVIN

NOISY Animal ABC
Frances Mackay

Animal ABC Activity Book
COLOURING   MATCHING
LETTER FORMATION
PUZZLES
WRITING   DOT-TO-DOT
Ages 4-7
Frances Mackay

MONSTER COUNTING Activity Book
Learn to count to 20
NUMBER TRACING
COUNTING
MATCHING   PUZZLES
COLOURING   DOT-TO-DOT
WRITING
Ages 2-7
Frances Mackay

Dinosaur Activity Book
50 Fun things to do!
PUZZLES
COLOURING
WRITING
DOT-TO-DOT
DRAWING
Ages 4-9
Frances Mackay

My Feelings Activity Book
PUZZLES   POSTERS
DRAWING   COLOURING
WRITING
50 Fun things to do!
Ages 7-11
Frances Mackay

Mammals and Birds of TASMANIA
FRANCES MACKAY
With Fun Facts & Printable Activities

AWESOME FACTS About TASMANIA AUSTRALIA
WITH PRINTABLE ACTIVITIES
FRANCES MACKAY

DOGS Counting Activity Book
Learn to count to 20
Ages 2-7
Frances Mackay

Acknowledgements for Dogs, A Counting & Comparing Book
All illustrations are licensed from Dreamstime.com
Illustrators who have elements on these pages: page 5 Elena Stebakova, page 7 Mariia Mashkova, page 9 Oni Adhi and Alena Simonova, page 10 Chudtsankov, Page 11 Andrew Genn, Page 19 Yana Bolbot and Homoerectuss, page 21 Carbouval, page 23 Passengerz, page 24 Alena Simonova. All other illustrations - Igor Zakowski